COLOR BY NUMBER FOR KIDS:

MERMAID

COLORING ACTIVITY BOOK FOR KIDS

Color Test Page

Color Test Page

1-Black 2-Peach 3-Brown 4-Lime
5-Gold 6- Red 7-Sky blue

1-Black   2-Peach 3-Pink 4-Yellow
5-Green 6- Sky blue 7-Red

1-Black   2-White   3-Lime 4-Peach
5-Purple 6- Gold   7-Sky blue

1-Black   2-White   3-Green 4-Yellow
5-Silver   6-Orange   7-Sky blue

1-Black 2-Lime 3-Tan 4-Purple
5-Magenta 6-Sky blue 7-Silver

1-Black   2-White   3-Silver   4-Grey
5-Lime   6-Orange   7-Tan

1-Black 2-White 3-Silver 4-Gold
5-Lime 6-Tan 7-Red

1-Black 2-Grey 3-Lime 4-Sky blue
5-Red 6-Orange 7-Peach

1-Black   2-Grey   3-Silver   4-Sky blue
5-Tan 6-Purple 7-Peach

1-Black   2-Green   3-Grey   4-Sky blue
5-Peach   6-Lime   7-Tan

1-Black 2-Tan 3-Lime 4-Pink
5-Peach 6-Purple 7-Yellow

1-Black 2-Orange 3-Yellow 4-Peach
5-Lime 6-Pink 7-Violet

1-Black 2-Orange 3-Yellow 4-Peach
5-Lime 6-Pink 7-Violet

1-Black   2-Tan   3-Lime   4-Orange
5-Peach   6-Red   7-Sky blue

1-Black   2-Pink   3-Sky blue 4-Tan
5-Orange 6-Lime 7-Red

1-Black   2-Red   3-Sky blue   4-Peach
5-Orange   6-Lime   7-Green

1-Black   2-Red   3-Orange   4-Tan
5-Green   6-Yellow   7-Pink

1-Black 2-Red 3-Tan 4-Peach
5-Lime 6-Yellow 7-Gold

1-Black 2-Peach 3-Tan 4-Orange
5-Lime 6-Gold 7-Sky blue

1-Black   2-Tan   3-Lime   4-Sky blue
5-Gold   6-Green   7-Pink

1-Black 2-Peach 3-Tan 4-Orange
5-Gold 6-Lime 7-Sky blue

1-Black   2-Peach   3-Tan   4-Pink
5-Purple   6-Lime   7-Sky blue

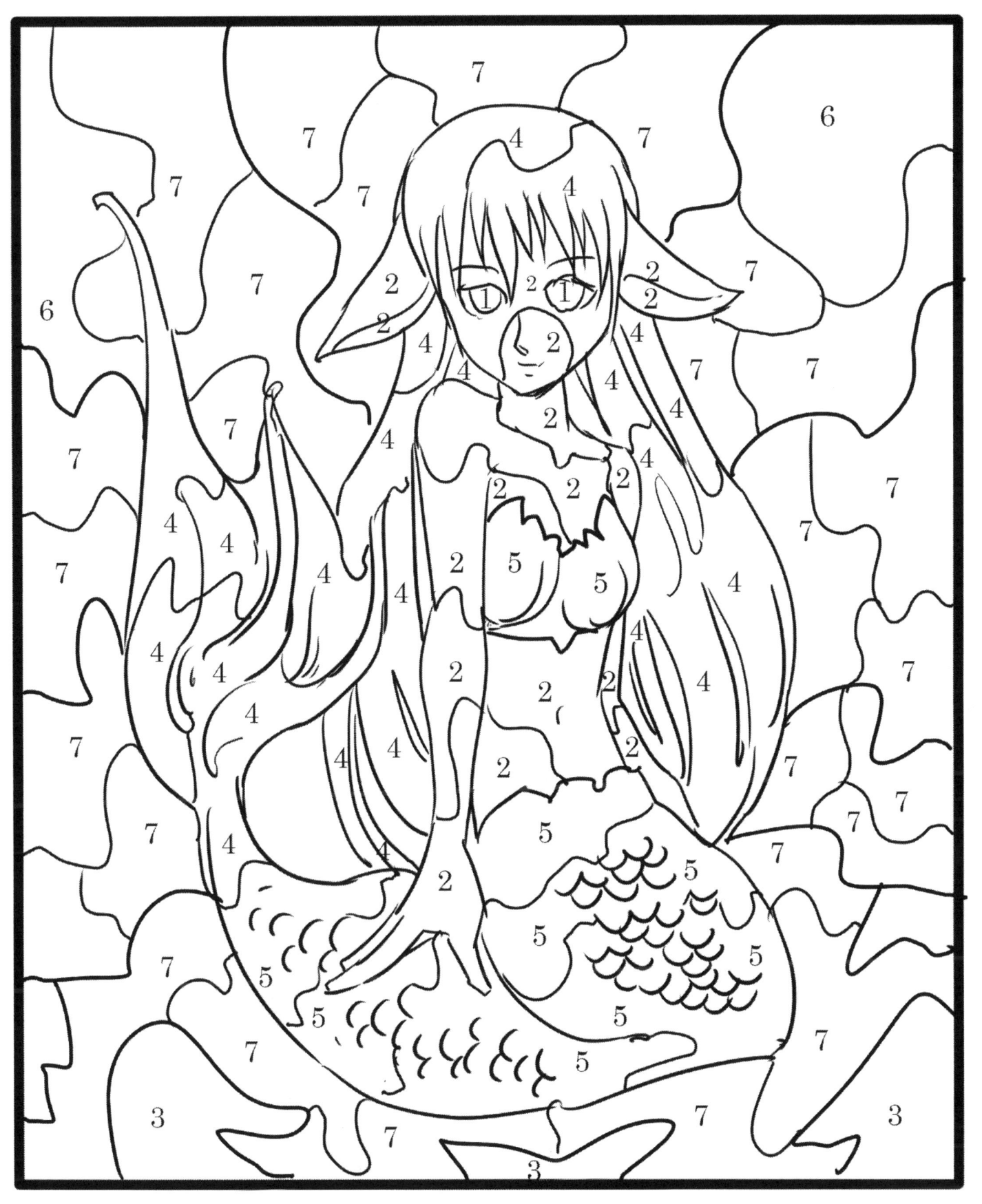

1-Black 2-Peach 3-Orange 4-Red
5-Gold 6-Lime 7-Sky blue

1-Black   2-Tan   3-Gold 4-Lime
5-Magenta 6-Grey 7-Sky blue

1-Black 2-Tan 3-Gold 4-Red
5-Sky blue 6-Lime 7-Yellow

1-Black 2-Peach 3-Red 4-Lime
5-Green 6-Pink 7-Sky blue

1-Black 2-Lime 3-Red 4-Peach

5-Green 6-Pink 7-Sky blue

1-Black   2-Peach   3-Orange   4-Tan
5-Yellow   6-Magenta   7-Sky blue

1-Black 2-Yellow 3-Lime 4-Peach
5-Gold 6-Pink 7-Blue

1-Black 2-Orange 3-Purple 4-Blue
5-Peach 6-Gold 7-Lime

Bonus